Awesome, Disgusting Careers

DISGUSTING SCIENCE JOBS

Mary E. Bleckwehl

BLACK RABBIT BOOKS

Hi Jinx is published by Black Rabbit Books
P.O. Box 227, Mankato, Minnesota, 56002.
www.blackrabbitbooks.com

Marysa Storm, editor; Michael Sellner, designer and photo researcher

Names: Bleckwehl, Mary Evanson, author.
Title: Disgusting science jobs / by Mary E. Bleckwehl.
Description: Mankato, Minnesota : Black Rabbit Books, [2023] | Series: Hi jinx. Awesome, disgusting careers | Includes bibliographical references and index. | Audience: Ages: 8-12 | Audience: Grades: 4-6 | Summary: "Let readers explore the awesome, disgusting science jobs that keep their world running smoothly through witty, conversational text, fun facts, and critical thinking questions that'll have them laughing and learning"- Provided by publisher.
Identifiers: LCCN 2020030451 (print) | LCCN 2020030452 (ebook) | ISBN 9781623106843 (hardcover) | ISBN 9781644665510 (paperback) | ISBN 9781623106904 (ebook)
Subjects: LCSH: Science-Vocational guidance-Juvenile literature. | Scientists-Juvenile literature
Classification: LCC Q147 .B53 2022 (print) | LCC Q147 (ebook) | DDC 502.3-dc23
LC record available at https://lccn.loc.gov/2020030451
LC ebook record available at https://lccn.loc.gov/2020030452

Image Credits

Alamy: Chris Howes/Wild Places Photography, 16–17; EyeEm, 10–11; Reinhard Dirscherl, 19; iStock: Pcanzo, 23; Science Source: Jim Varney, 12; Richard T. Nowitz, 12; Shutterstock: Alica in Wonderland, Cover, 11, 21; Angeliki Vel, 14; AnggaR3ind, Cover, 4; Arcady, 8; benchart, Cover, 4; BlueRingMedia, 4, 9; DenisNata, 7; DVitaliy, 20; George Rudy, 2, 3; Gorb Andrii, 15; Krisana Antharith, 12; LoopAll, Cover, 1, 5, 6, 20; Memo Angeles, Cover, 4, 14, 16, 17, 19; Midorie, 4; Nick Starichenko, 8–9; Pasko Maksim, 6, 23, 24; picoStudio, 13, 17, 18; Pitju, 21; Rich Carey, 19; ridjam, 12; RJC Cartoons, 6–7; Rohit Dhanaji Shinde, 2, 3; Ron Dale, 5, 6, 20; SatawatK, 7; Sujono sujono, 10, 21; Svetlana777, 12–13; totallypic, 10

CONTENTS

Chapter 1

SLIMY, SMELLY SCIENCE

Working in science means exploring weird ideas. Scientists learn how things work. They make life-changing discoveries. Some science jobs can get nasty, though. They can be slimy and smelly. Get ready to check out gross jobs that make a difference.

Chapter 2

THE JOBS

Pet Food Tester

You might eat pet food on a dare. But some people do it for a job. First, these testers sniff the food. Then, they pop it in their mouths. They write down how the food tastes and feels. Food that's no good goes back to the drawing board. These workers make sure pets get tasty, **nutritious** chow.

Most testers don't swallow pet food. They spit it out.

pet food testing

For about 40 years, George Aldrich has been test smelling products for NASA. In space, there's no way for smells to escape. Aldrich makes sure astronauts won't be stuck with gross smells.

Professional Smellers

The nose knows! Professional smellers put their noses to work. But that means sticking them in some gross places. Some sniff armpits. They check if new deodorant is working. Others take whiffs of kitty litter. They make sure it's getting rid of bad smells.

Insect Scientist

Some science jobs have people working with bugs. Others have people making food. Some jobs combine the two. That's right. Some insect scientists turn creepy crawlies into food! How would you feel about eating insect ice cream? That's just one food these scientists have created.

insect ice cream

forensic entomologists

bugs

crime scene

Forensic Entomologist

Some scientists use bugs to solve crimes. Bugs often show up to eat dead bodies. When police find bodies, they call forensic entomologists. These scientists know all about bugs and their life cycles. They study the bugs crawling around dead bodies. They look at the bugs' sizes and ages. All this information helps tell when people died.

Some doctors use maggots. They have the little worms eat dead **tissue** on a person's wound.

Collecting Animal Pee

Scientists do a lot of gross things. And collecting animal pee is one of them. They don't chase animals all day, though. Instead, they use their creativity and a few tools. Sometimes they collect pee from pens with slanted floors. All the pee flows to one end. Others gather pee that drips through grated floors. Collecting pee is only half of this job, though. Scientists then study the pee. They make sure the animals are healthy.

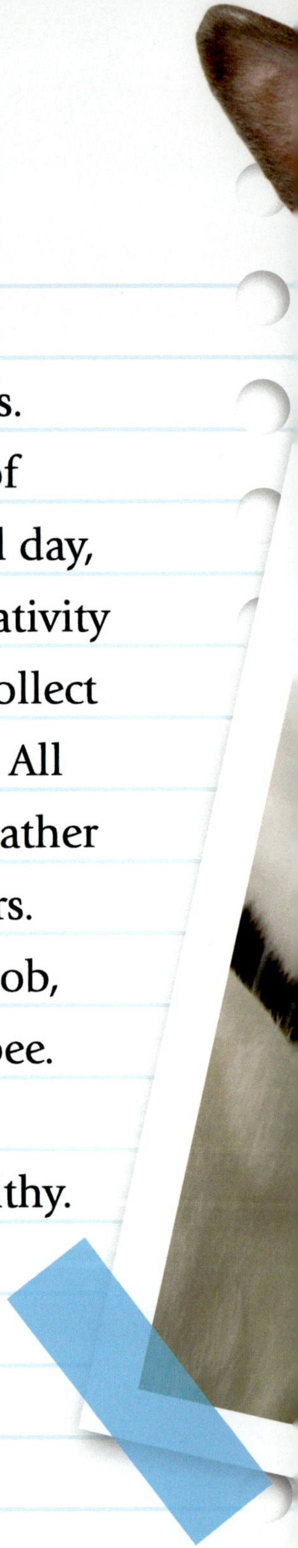

Animal pee has uses! Hunters often use female deer pee to **attract** bucks.

Bat Poop Researcher

You might want to plug your nose for this one. Some scientists study bat poop. This poop is found in caves. It's slippery, stinky, and often knee-deep. Researchers must wear tall boots and headlamps. These workers scoop up samples to study. Their studies tell them about bats' **diet** and habitat health.

Oceanic Snot Diver

See that scientist diving into the Mediterranean Sea? He isn't treasure hunting. He's swimming through marine **mucilage**. These smelly snot-like globs form when dead and living matter join together. These blobs hurt ocean life. Scientists collect samples of the snot. They study them under microscopes. They also perform tests on them. Learning about this snot can someday help scientists get rid of it.

This slime hurts fish by coating their gills. These fish can't breathe.

Chapter 3

GET IN ON THE HI JINX

People working science jobs need more than a strong stomach. Most also need a college education. They must take math and science classes. They also need curious minds. If you like to learn, a science job might be for you. You can start your path now by asking questions. Head to the library, and learn about the world.

Take It One Step More

1. Which job in this book would you never do? Why?

2. What other **traits** does a person need for a science job?

3. Would you ever eat food made from bugs? Why or why not?

GLOSSARY

attract (uh-TRAKT)—to draw someone or something to oneself

diet (DYE-uht)—food and drink an animal eats

gill (GYL)—an organ in fish used to get oxygen from water

mucilage (MYOO-suh-lij)—a jellylike substance of various plants

nutritious (noo-TRISH-shus)—having substances that a person or animal needs to be healthy

tissue (TIH-shu)—a mass of cells enclosed with another substance that forms the basic structural materials of plants and animals

trait (TREYT)—a characteristic or quality

LEARN MORE

BOOKS

Mattern, Joanne. *Disgusting Jobs.* That's Disgusting! Minneapolis: Kaleidoscope, 2020.

Nickel, Scott. *World's Grossest Jobs.* Minneapolis: Lerner Publications, 2021.

Walker, Kevin. *STEAM Jobs in Agriculture and Food Development.* STEAM Jobs You'll Love. North Mankato, MN: Rourke Educational Media, 2020.

WEBSITES

13 Facts to Gross Out Your Parents
kids.nationalgeographic.com/explore/13-facts-to-gross-out-your-parents/

Fun Facts about Bugs
www.si.edu/spotlight/buginfo/fun-facts-bugs

Guano Facts for Kids
kids.kiddle.co/guano

INDEX